FIRST OF ALL, THANK YOU!

Thank you for trusting us by choosing our books.
We hope that you will enjoy this book! If so, please leave a positive
review, it motivates us to create better books.

Doodle Noodle

Enter the realm of Doodle Noodle, where imagination dances with
hues and pigments come alive! Our enchanting coloring book brand
invites you to experience a symphony of colors and embark on a
whimsical journey of self-expression. With Doodle Noodle, you have
the power to create extraordinary compositions, infusing life and
magic into every stroke. Unleash your creativity, indulge in the joy of
coloring, and let your dreams radiate with vivid brilliance!

Doodle ♥ Noodle

THIS BOOK BELONGS TO:

....

....

....

....

COLOR TEST PAGE

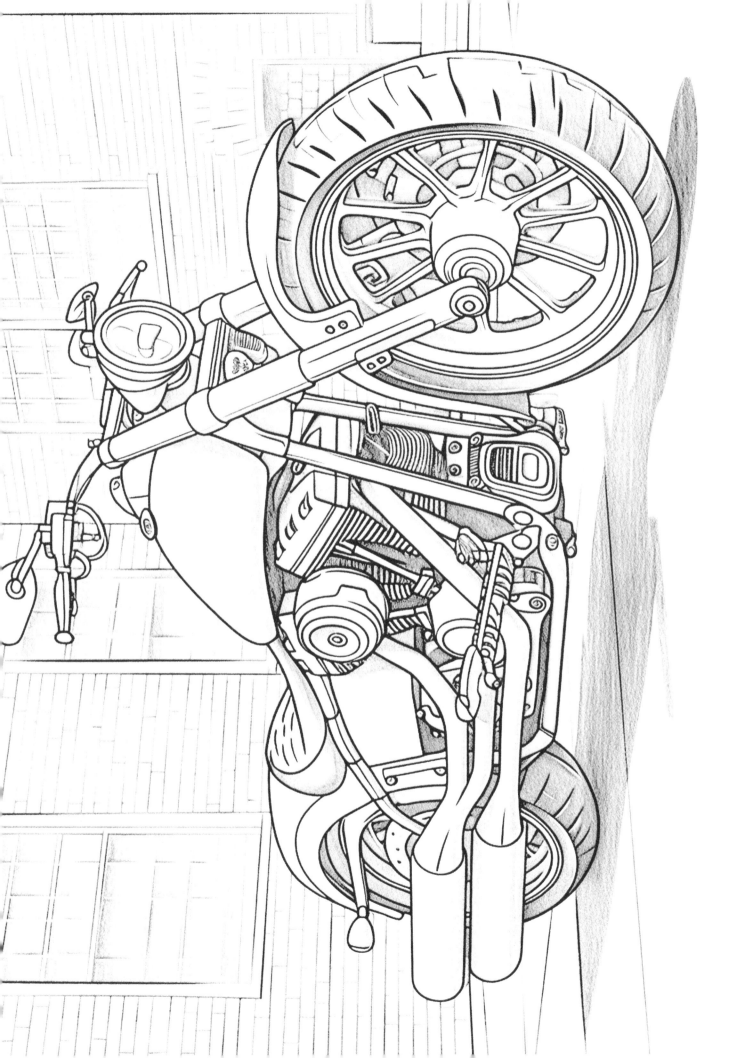

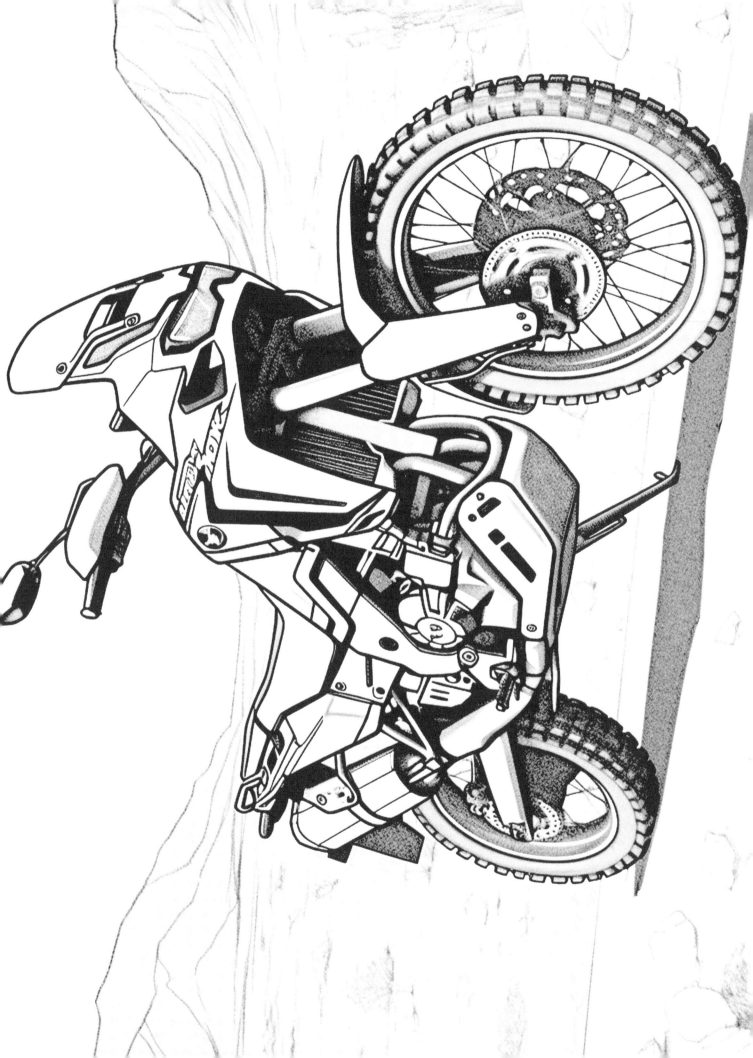

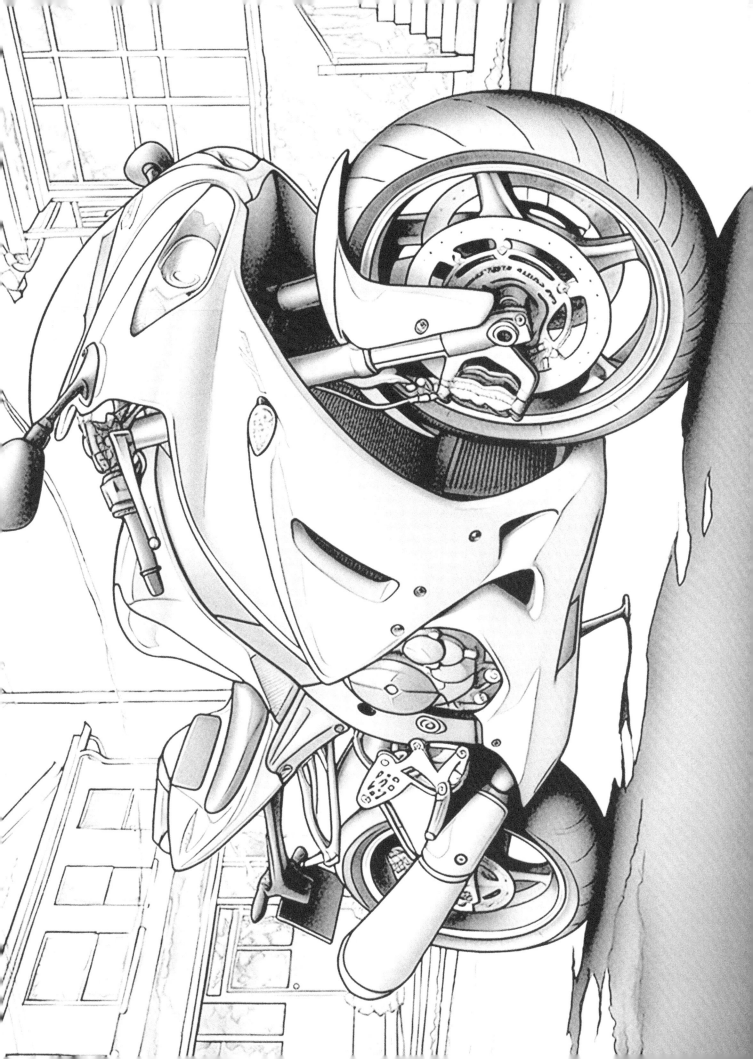

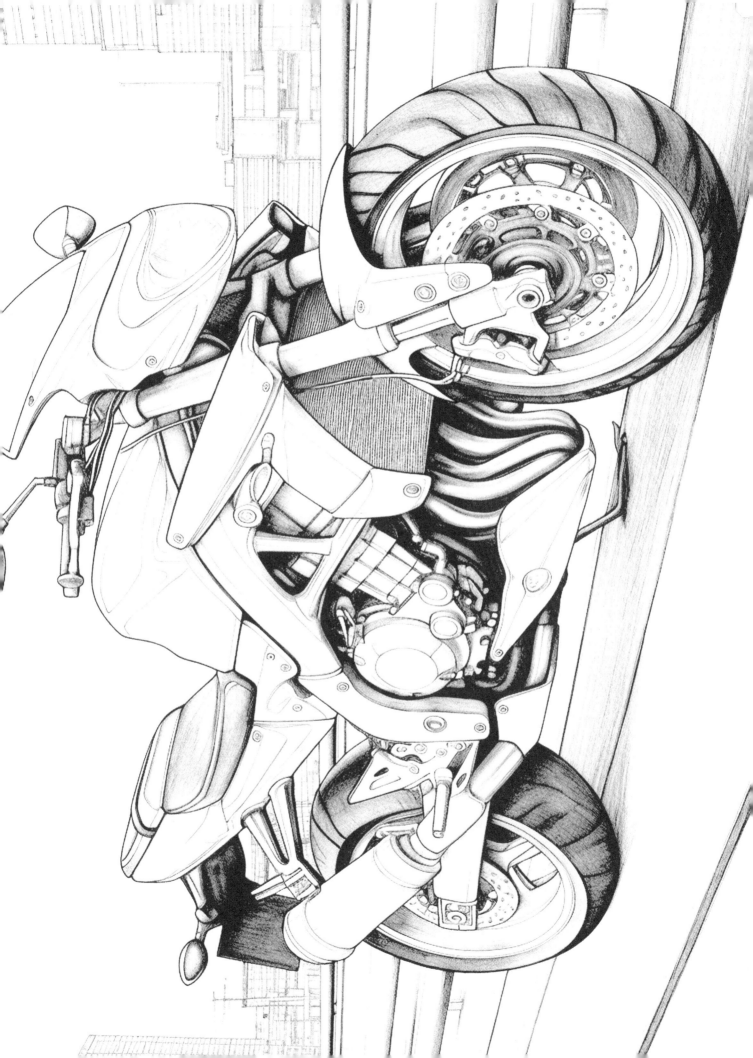

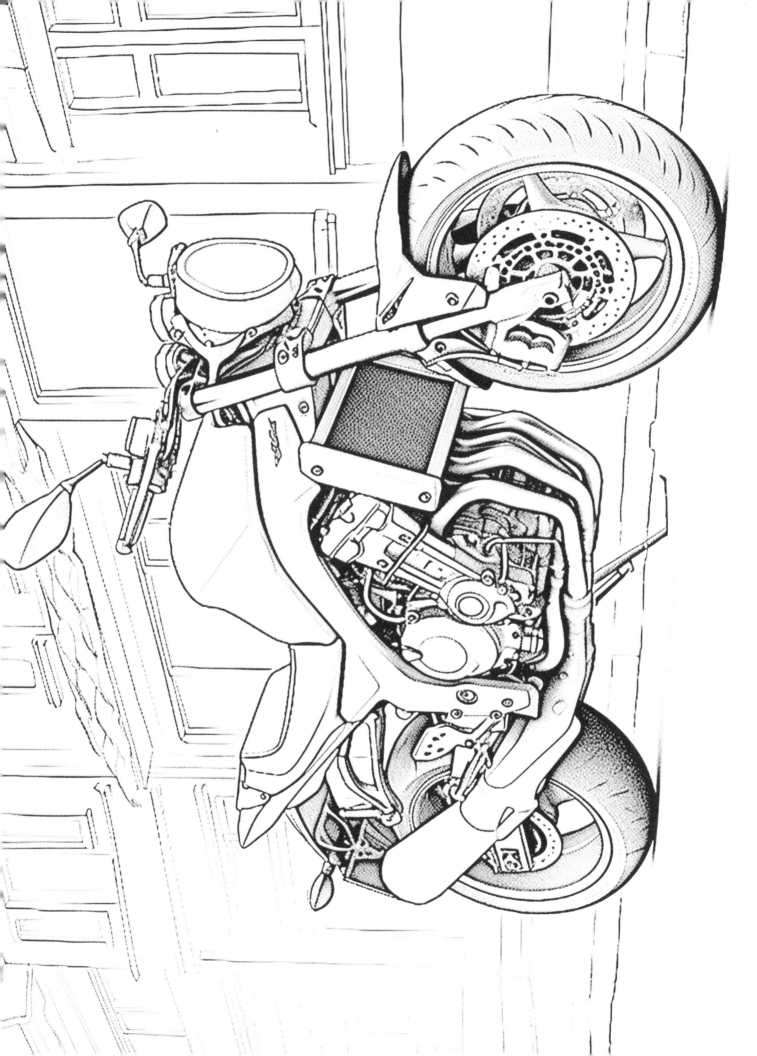

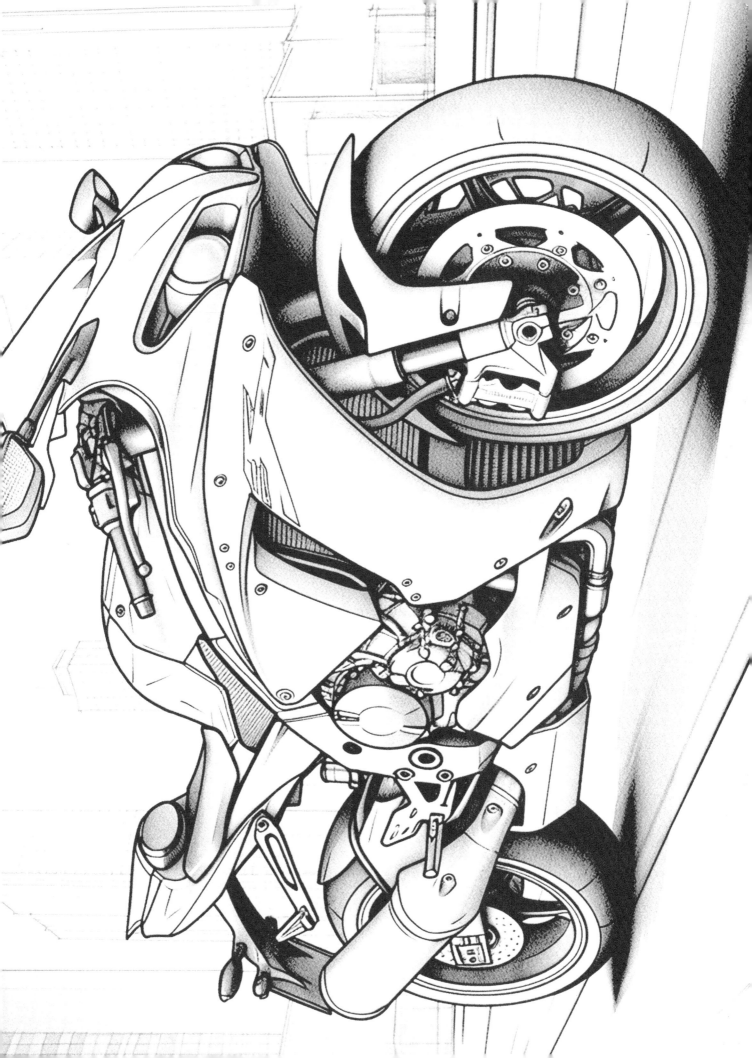

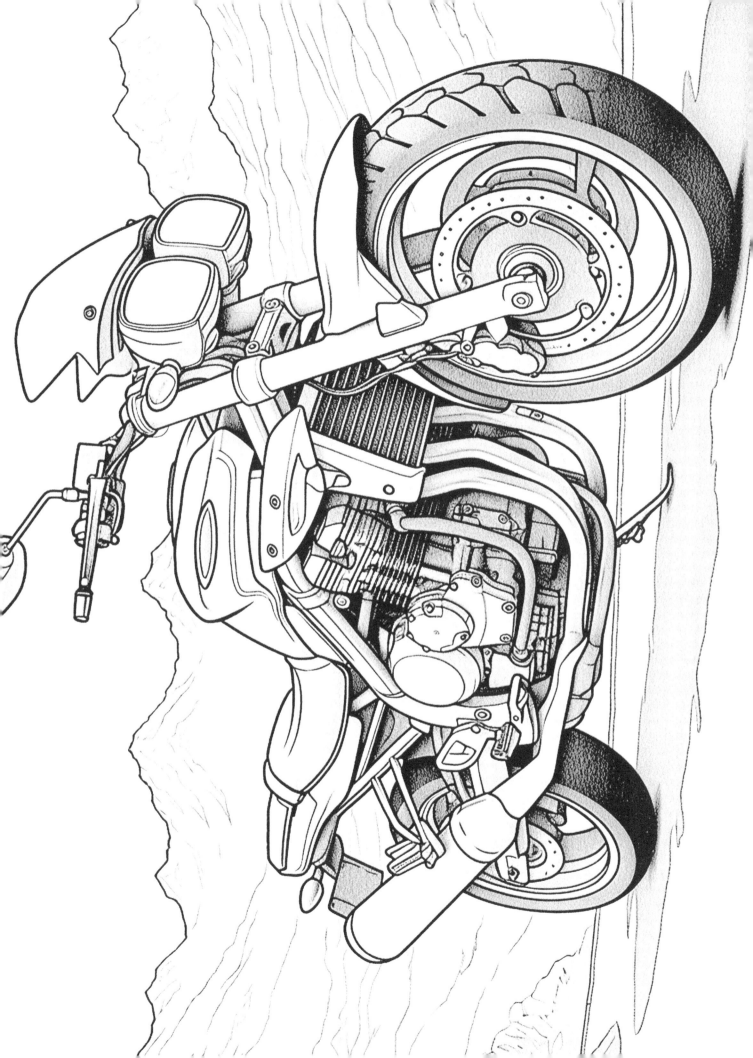

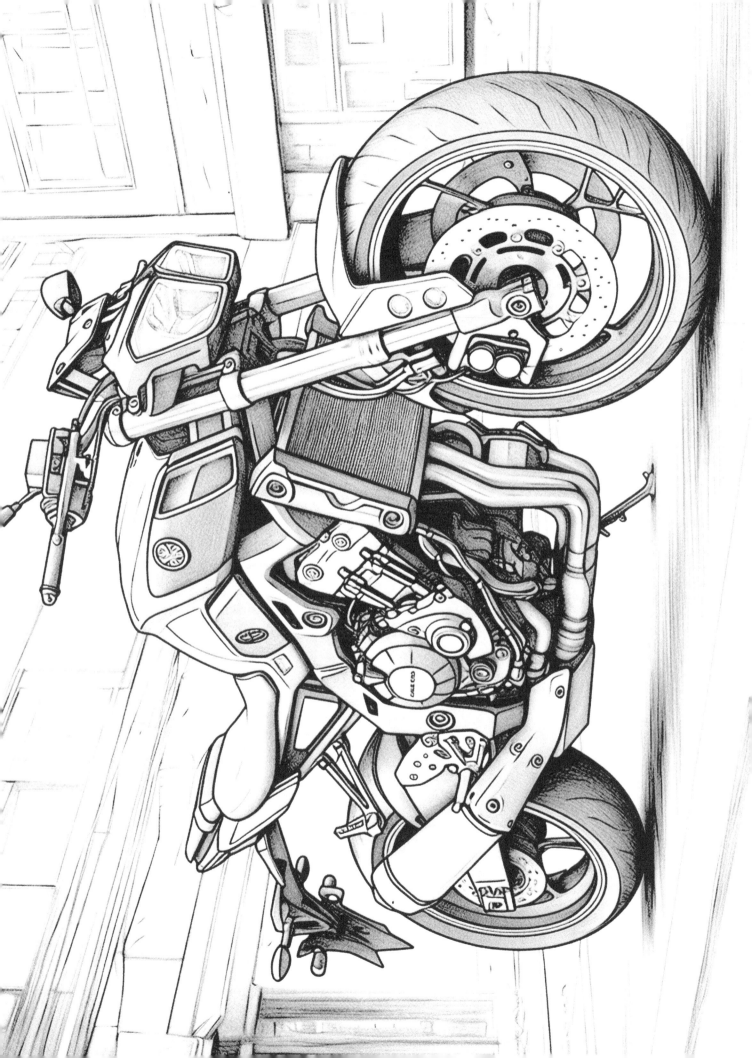

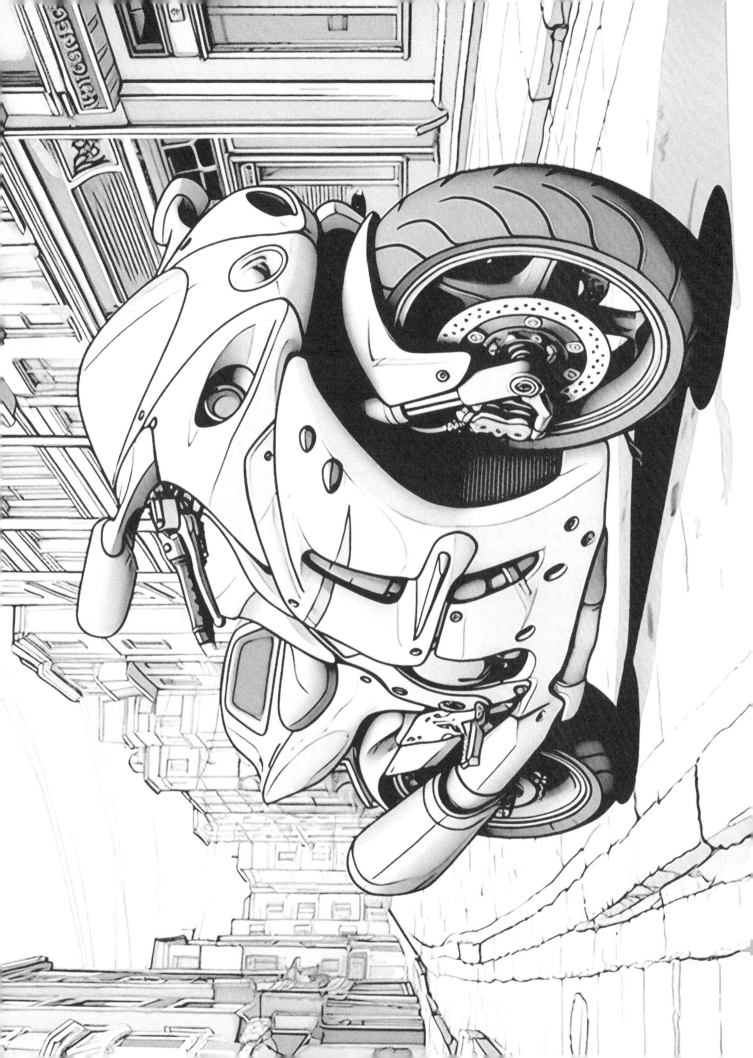

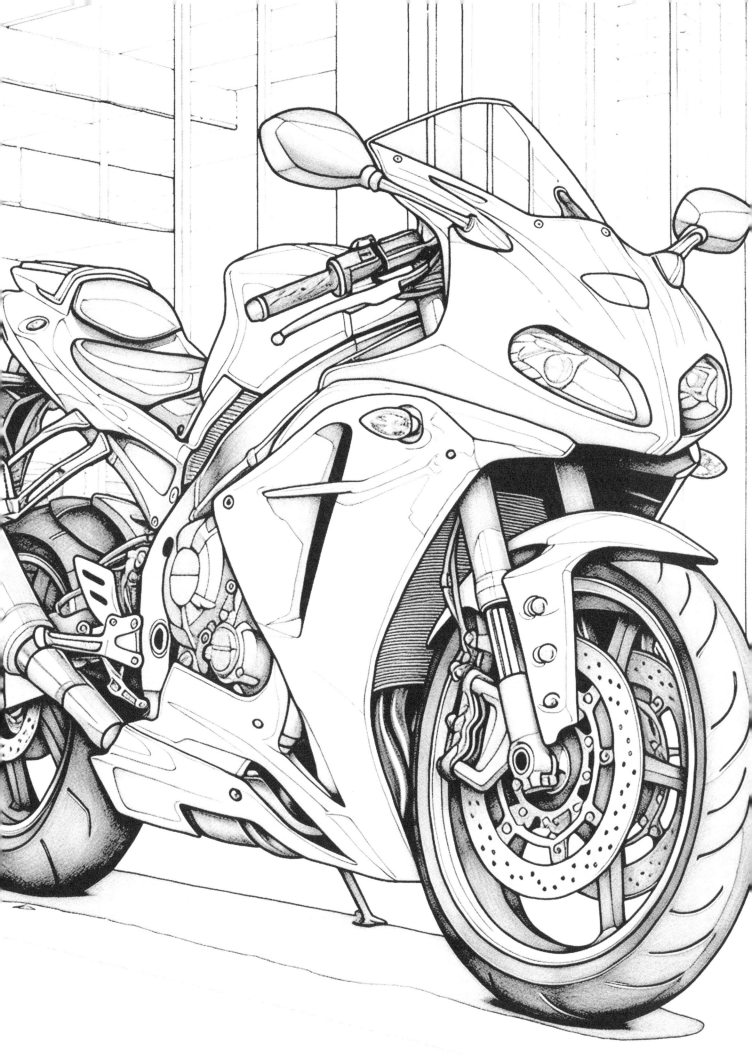

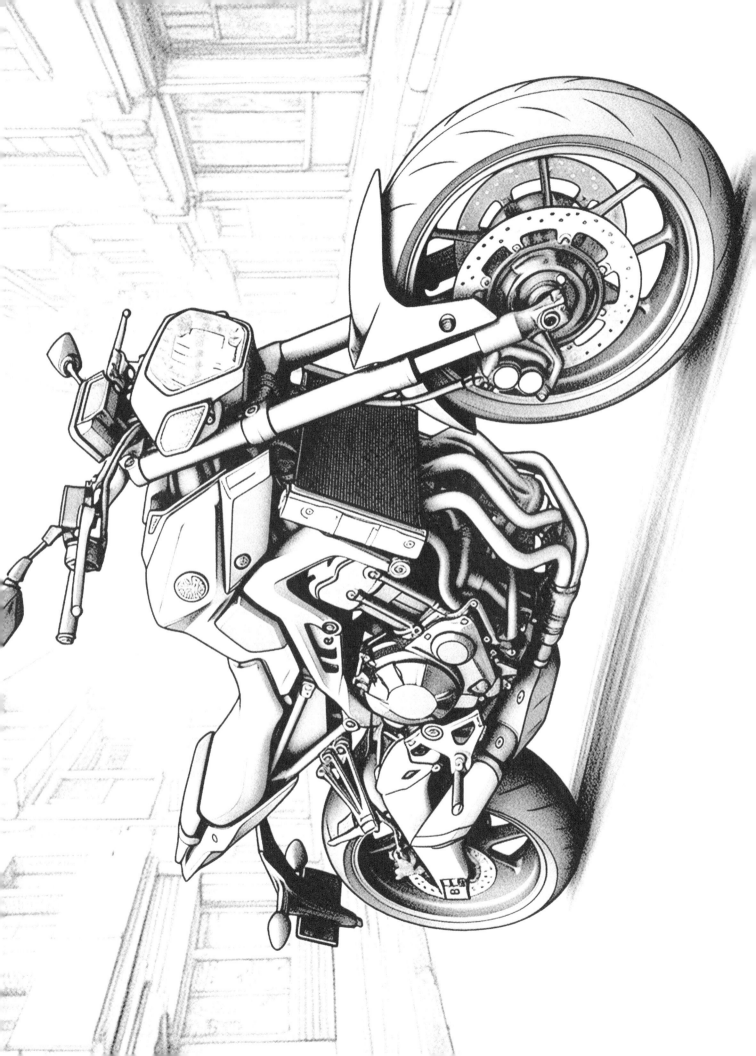

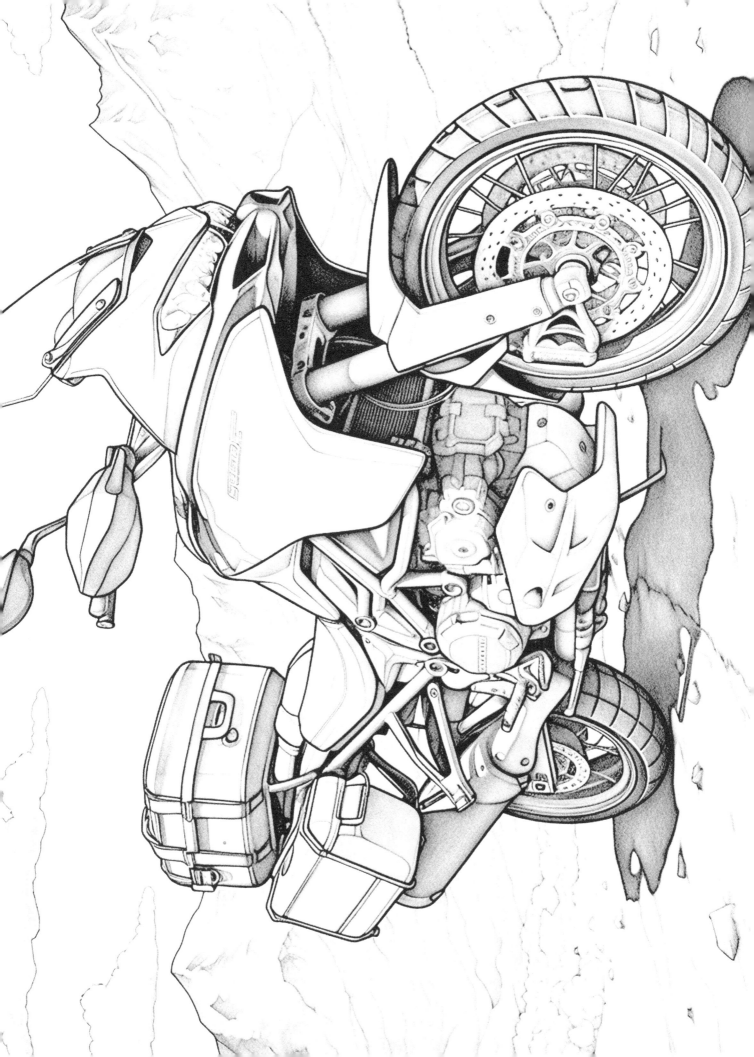

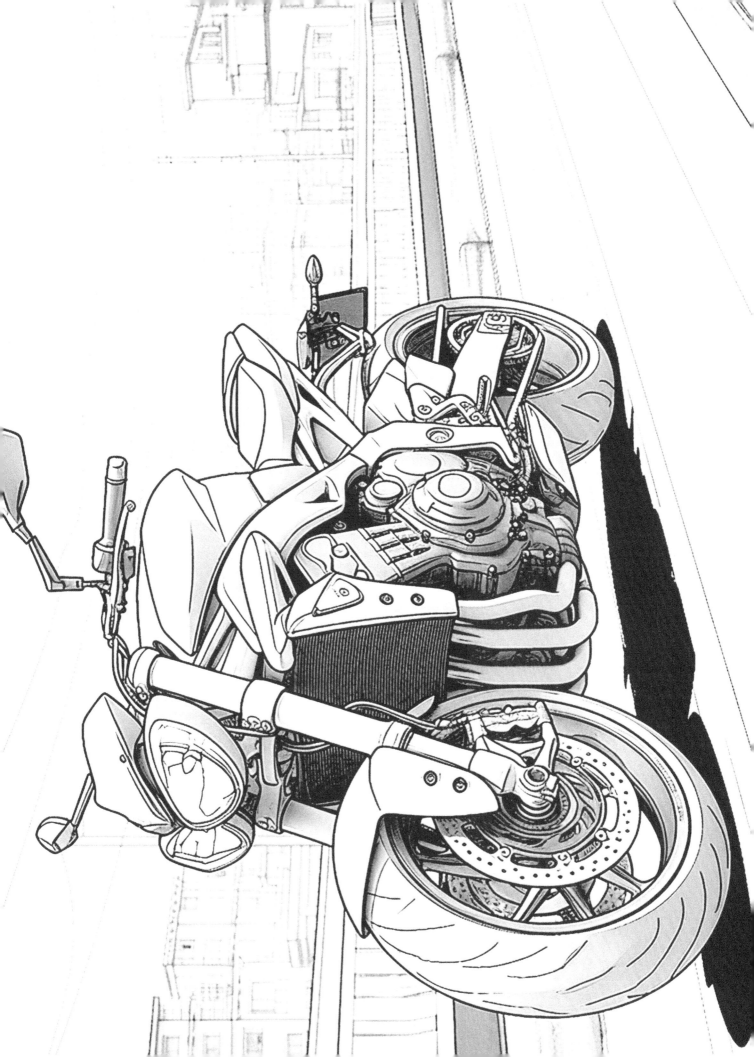

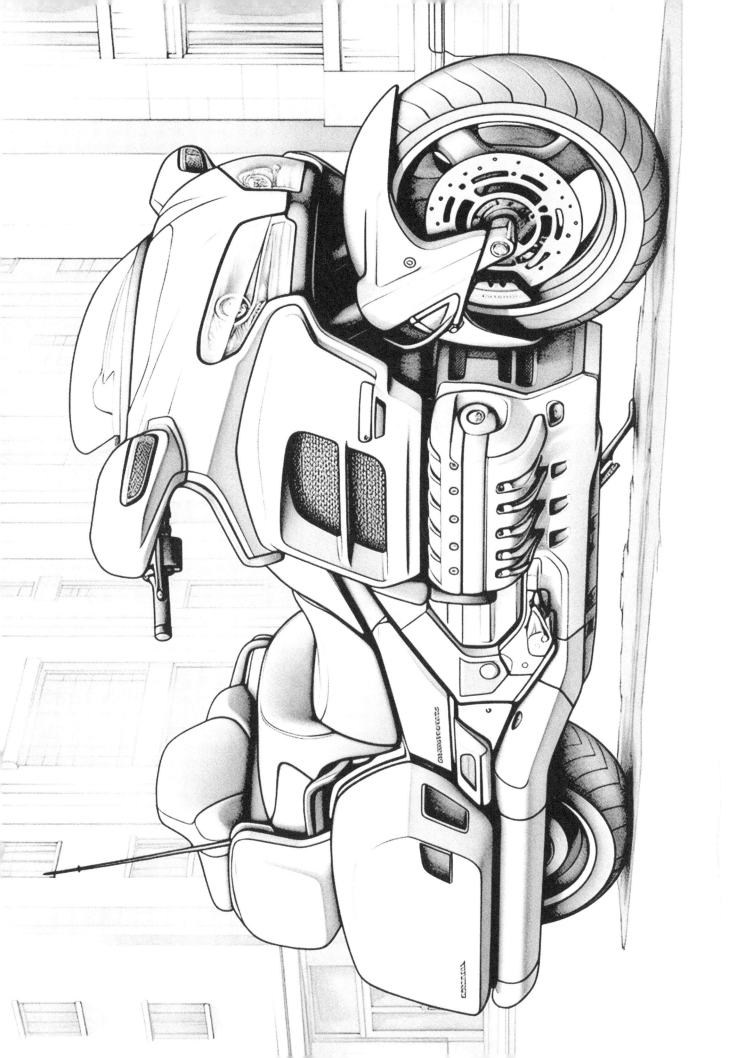

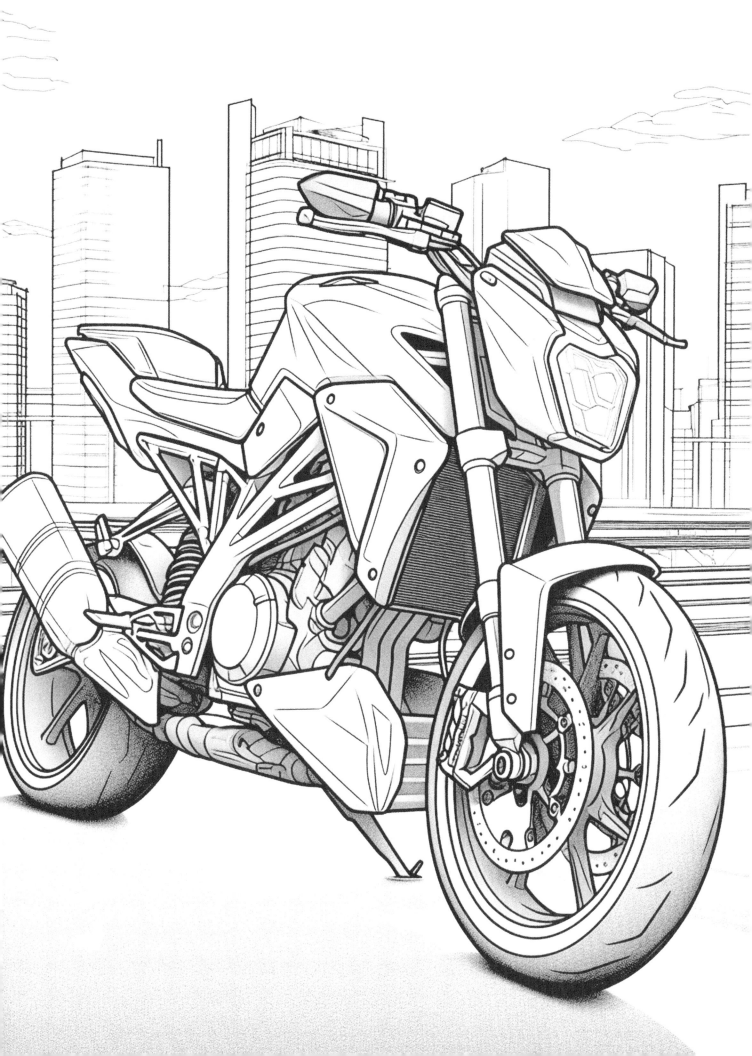

WRITE DOWN YOUR FAVORITE ASPECTS
OF THIS BOOK:

Doodle Noodle

THANK YOU FOR TRUSTING US BY CHOOSING OUR BOOKS.

Your trust in us means a lot, and we truly hope that you will find joy and satisfaction in coloring our unique designs. If our book meets your expectations, we kindly ask you to leave a positive review as it motivates us to create even better books in the future. Once again, thank you for your support and we hope that our coloring book will bring a little bit of creativity and relaxation into your life.

Made in the USA
Monee, IL
05 December 2023